A COLLECTION OF COCKROACHES

BY REBECCA STORM

CONTENTS

COCKROACHES	4
COCKROACHES UP CLOSE	6
UNFUSSY EATERS	8
ON THE MOVE	10
SEARCHING FOR FOOD	12
STINKY INSECTS	14
FINDING A MATE	16
LIFE CYCLE	18
NYMPHS GROWING UP	20
HUMANS AND COCKROACHES	22
ODD BEHAVIOUR	24
ALL SHAPES AND SIZES	26
FUN COCKROACH FACTS	28
GLOSSARY	30
INDEX	32

Copyright © 2025 Hungry Tomato Ltd

First published in 2025 by Hungry Tomato Ltd
F15, Old Bakery Studios, Blewetts Wharf, Malpas Road, Truro, Cornwall, TR1 1QH, UK.

No part of this publication may be reproduced, stored in a retrieval system, or transmitted in any form or by any means, electronic, mechanical, photocopying, recording, or otherwise, without prior written permission of the copyright owner.

A CIP catalogue record for this book is available from the British Library.

ISBN 9781835694152

Printed in China

Discover more at www.hungrytomato.com

Picture credits:
Abbreviations: m-middle, t-top, l-left, r-right, bg-background.

Getty Images 11br. Nature Picture Library: 8b. Premaphotos wildlife: 26mr, 27bl. Science Photo Library: 27t. Shutterstock: 6-7bg, FC, 9b, 17b, 21tr, 25br; 19 STUDIO 16m, 31br; anto.cakep 15t; Chris Ne 5tr; BANDZRIO 3bl, 12ml; Bhitakbongse Leesothikul 9tr; BorneoJC James 29tr; Dr.Morley Read 20b; guentermanaus 15b; Guillermo Guerao Serra 23tl; Holger Kirk 13t; Jpreat 10b; Kurit afshen 19br; Liz Weber 23ml, 28mr; Lukas Jonaitis 7tr; Mr.Pattrawut Yamyeungyong 4tl, 11tr, 28bl; NattapolstudiO 19bl; Nick Greaves 21br; NuayLub 19ml; Salouw 4b; Tomasz Klejdysz 18b; Usa.P 14mr; RHJPhotos 1bg, 5br, 17tr, 22br, 23bl, 23br; Ridholaresha 13br; Vera Larina 29ml.

Every effort has been made to trace the copyright holders, and we apologise in advance for any unintentional omissions. We would be pleased to insert the appropriate acknowledgements in any subsequent edition of this publication.

DISCLAIMER:
Insects are fascinating, but best to stay away! Don't touch or handle them – some insects can sting or get aggressive when they feel threatened.

Words in **BOLD** can be found in the glossary.

COCKROACHES

Cockroaches are winged **insects**. They like warm, damp, dirty places. Cockroaches are very tough creatures that do not squash easily. They can run very quickly too!

HOW DO THEY LIVE?

Cockroaches either live alone or in small family groups consisting of a female and her young. They are normally only seen at night.

WHERE DO THEY LIVE?

Cockroaches like to live in warm, damp conditions. Most cockroaches live in **tropical** and **sub-tropical** forests. However, some have invaded human settlements and are now found in cities throughout the world.

A forest cockroach in its leafy forest home

WHAT DO THEY EAT?

Cockroaches have big appetites – they will feed on just about any plant or animal, as long as it is already dead! Animals that eat like this are known as **scavengers**.

A death's head female cockroach with her young

IT'S A BUGS WORLD

Insects belong to a group of animals known as **arthropods**. Adult arthropods have jointed legs, but do not have an inner **skeleton** made of bones. Instead, they have a tough outer "skin" called an **exoskeleton**. Most insects have at least one pair of wings.

Cockroaches have wings, but they can't fly very well.

COCKROACHES UP CLOSE

The average cockroach is around 4 cm (1.5 inches) long and is reddish-brown in colour. They have a tough outer covering that gives them a smooth, shiny appearance.

Beneath this covering, the cockroach has the same sort of body as all other adult insects. It is divided into three parts – head, **thorax**, and **abdomen**.

The abdomen is the largest part of the cockroach's body. It contains the **digestive system**, where the cockroach processes its food.

The thorax is the middle part of the body. The legs and wings are attached here.

SUPER SIX LEGS

Beetles and other insects are sometimes called "hexapods" because they all have six legs ("hex" means "six" in Greek). This can be a bit confusing – all insects are hexapods, but not all hexapods are insects! Some bugs, such as **springtails**, have six legs but they are not true insects.

Springtail

The head has **antennae**, eyes, a mouth and part of the brain. The head is often hidden from sight beneath a protective shield called a **pronotum**.

UNFUSSY EATERS

Cockroaches are not carnivores or herbivores. Instead, these bugs are omnivores, which means that they will eat any kind of plant or animal food – but only if it is already dead!

Cockroaches are not **predators** that hunt and kill other animals for food. Cockroaches are scavengers. They prefer their food to not just be dead, but **rotting**!

Cockroach eating dead leaves

Animals and plants gradually turn soft and mushy as they **decompose**. Most cockroaches will also feed happily on animal droppings! Cockroaches are part of nature's clean-up crew.

A cockroach eating animal droppings

LIVING THE LOG LIFE

Some cockroaches, known as wood roaches, love to feed on fallen trees. Wood, even rotten wood, is a very poor quality food. Wood roaches are able to live on this diet thanks to a **microbe** that lives in their digestive system. As a result, wood roaches do not have to search of food – they can spend all their time safely inside a single rotten log!

Wood roaches sitting on a dead log

ON THE MOVE

During the daytime, cockroaches hide from predators. Their flat bodies help them squeeze into the smallest hiding places, where they wait for darkness.

At night, the cockroaches come out to feed. Although there are fewer predators around, the roaches move very quickly just in case. When it comes to walking and running, cockroaches are pretty speedy!

Cockroaches walk in the same way as other insects – they lift the middle leg on one side at the same time as the front and back legs on the opposite side. This means there is always three legs touching the ground, making them very stable and unlikely to fall over.

SPRINTING EXPERTS

When cockroaches run, they really run! They lean back and lift the front of their bodies into the air so that they end up **sprinting** on just their back legs! At full speed, some cockroaches can cover up to 50 body lengths per second – that's about 10 times faster than a human runner!

SEARCHING FOR FOOD

Cockroaches have very poor eyesight, and most can do little more than tell the difference between light and dark. But they make up for this by having long and sensitive antennae.

The antennae are very flexible because they are divided into around 100 **segments**. Each segment carries sensitive **receptors** that each cockroach uses to find out about its surroundings.

Some of the receptors are sensitive to **vibrations**. Other receptors are sensitive to temperature – there are separate receptors for hot and cold.

The most important receptors are those that allow cockroaches to sample smells. Different receptors sense different smells, especially the smells made by rotting things!

A cockroach eating a rotten banana

SUPER SENSITIVE

As well as their antennae, cockroaches can also detect vibrations through tiny **bristles** on their legs. Even while their antennae are busy finding food, their legs are alert to the slightest movements around them.

The bristles on a cockroach's leg can detect the slightest movement.

STINKY INSECTS

Cockroaches are not nice to be around! Mostly because they have a horrid smell. However bad the smells of rotting meat or plants are, cockroaches can smell even worse!

BAD HYGIENE

What makes cockroaches really unpleasant is that they have no toilet training. They leave a non-stop trail of droppings wherever they walk! Because their mouths point backwards, cockroaches have to walk over their food in order to eat it. They even leave droppings on food they have not eaten.

Cockroach poop droppings

Some cockroaches use their smell to protect themselves. If threatened by a predator, they can release a horrible-smelling liquid. As the predator moves away from the bad smell, the cockroaches can escape.

A scorpion and cockroach encounter

The message to other animals is very clear – "Avoid this unpleasant insect." The bad smell a cockroach makes might be a message to other cockroaches saying, "Here is plenty of food." Or, it can have the opposite meaning, "Keep away, this food is mine!"

FINDING A MATE

Both male and female cockroaches have wings, but most female cockroaches cannot fly. Only males have wings that can be used for flying, and they mainly use them for finding a **mate**.

Most of the time, male and female cockroaches live completely separate lives. When it is time to mate, the females release special substances called **pheromones** into the air, which get carried away with the wind.

FEMALE COCKROACHES ONLY

Not all bugs have to mate to produce babies. Some insect **species**, such as the Surinam roach, are all female and they can produce eggs without the need for any males!

Once the female pheromones reach a male, he takes to the air and begins following the invisible pheromone trail back towards the female.

If the male cannot find the female straight away, he lands and releases his own pheromones. They're not as strong, but they act in the same way, and help the female find the male in the darkness.

A male cockroach in flight

Surinam cockroach

LIFE CYCLE

There are over 4,000 types of cockroaches. After mating, most female cockroaches lay their eggs inside a special case called an ootheca. Depending on the type of cockroach, there can be up to 50 eggs neatly arranged inside.

Some types of cockroaches carry the ootheca around with them for a few days, before leaving it in a suitably dark and damp place. Other kinds of cockroaches, however, continue to carry it for a few weeks until their eggs hatch.

A female cockroach laying an ootheca

All oothecae have an outer surface which is smooth and leathery. There are tiny openings at the bottom to allow the eggs to breathe.

Only a few types of cockroaches do not lay eggs. Instead, the eggs develop inside the female's abdomen. The young cockroaches, called **nymphs**, are born live.

INSECT DEVELOPMENT

Insects can develop from eggs in two different ways. With many kinds of insect, including cockroaches and grasshoppers, the eggs hatch into nymphs that already have the adult body shape. However, with other kinds of insect, such as bees and beetles, the eggs hatch into **larvae** that look very different from the adults. Larvae go through a stage called **pupation** when they change into adults.

Wood cockroach nymph

Cockroach eggs

A horn beetle larvae

NYMPHS GROWING UP

Young cockroach nymphs look like small versions of the adults, but they are far from complete. It can take up to 10 months for nymphs to reach adulthood. During this time, the youngsters will shed their "skin" many times.

Insects have an exoskeleton that supports and protects their bodies. Exoskeletons are strong and tough, but do not stretch. In order for the insect to grow, it must grow a new exoskeleton before shedding the old one.

A cockroach shedding its old exoskeleton

The process of an animal shedding its outer covering is known as **moulting**. This term is not only used with insects and other arthropods, but also with some animals that have internal skeletons, such as snakes.

Newly hatched cockroach nymphs are pale and almost colourless, but they soon begin to turn darker. The nymphs will go through as many as 12 moults before they are fully-grown adults. The stages between each moult are known as **instars**.

The process of moulting

STAGED GROWTH

With each instar, the nymphs become more complete. For example, newly hatched nymphs have no wings at all, and their antennae have only 25 segments, compared with the 100 segments on adult antennae. Until the nymphs reach their final stage of growth, their wings and antennae will not be fully **functional**.

A tussock moth instar

HUMANS AND COCKROACHES

Cockroaches are most at home in warm, wet woodlands, but a long time ago they found out that human settlements were equally inviting...

Houses designed to be comfortable for people are also very comfortable for cockroaches! They provide everything the roaches need – heat, dampness, and large amounts of organic waste (unwanted bits of plants and animals).

Some cockroach species, around 20 in total, are so closely attached to human settlements that they have become serious pests. These cockroaches are now found in nearly every town and city in the world!

A cockroach infestation in a kitchen

Among the worst pests are the German cockroach (which actually come from Africa), the Oriental cockroach, and the American cockroach. These insects are often found living under floors and between walls, especially in bathrooms and kitchens.

German cockroach

Oriental cockroach

American cockroach

SPOILAGE

Cockroaches are pests not only because of the food they eat, but also because of the food they spoil! Their bad habits of crawling all over food and scattering droppings everywhere means they spoil a thousand times more food than they actually eat.

ODD BEHAVIOUR

Some cockroaches do some fairly odd things – well, odd for cockroaches! Not only can the largest species grow to an amazing 9 cm (3.5 inches) long, but the fastest cockroaches are sometimes entered into races by their human owners.

HISSING MONSTER

The Madagascan giant cockroach is not only one of the biggest roaches, but also the noisiest – so noisy, in fact, that it uses sound as a defensive weapon! When this cockroach is disturbed, it puffs its body up with air and then huffs out the air through tiny openings on its body. The air makes a loud hissing sound as it comes out, scary enough to warn off predators.

Madagascan giant cockroach

PARENTAL CARE

Most young cockroaches do not get any parental care. But this is not the case with wood roaches. When these nymphs hatch, they stay close to their mother and feed on her droppings. These droppings contain the microbes which will allow the roaches to feed on wood. Only after several weeks do the nymphs have enough of the microbes to start feeding on wood.

FAST RUNNERS

Many cockroaches can run fairly quickly, but none are faster than the American cockroach! This amazing insect can cover a distance of 150 cm (60 inches) in one second – about five times faster than a German cockroach, which can only cover about 30 cm (12 inches) in one second.

In some places, cockroach racing is a well-organised sport!

ALL SHAPES AND SIZES

Although most cockroaches have the same basic body parts, there is a lot of variety in shape, size, and colour.

LEAF COCKROACH

This West African cockroach hides from daytime predators by standing on the forest floor and keeping very still. It is perfectly **camouflaged** as a fallen yellow leaf, complete with one area that appears to be gradually turning brown.

A cockroach in disguise

BANANA ROACH

The green banana roach does not invade houses, but it is considered a pest because it feeds on crops.

Banana roach

TROPICAL GIANT COCKROACH

This cockroach from the tropical forests of Central America is one of the largest cockroaches in the world and, when fully grown, can measure more than 9 cm (3.5 inches) in length.

Tropical giant cockroach

Cape Mountain cockroach

CAPE MOUNTAIN COCKROACH

The Cape Mountain cockroach lives in Africa. Unlike most other cockroaches, the Cape Mountain cockroach does not lay its eggs in an ootheca. Instead, the eggs remain inside the female's body until the nymphs hatch.

FUN COCKROACH FACTS

There's so much more to know about cockroaches! Delve into some fantastic facts about these unusual creatures.

THE BRAIN OF A COCKROACH IS...

found along the underside of its belly. That's why if a cockroach's head is cut off, it can survive for up to a week!

THEY HAVE BEEN...

on Earth for more than 400 million years.

A COCKROACH'S MOUTH CAN...

smell as well as taste, and is able to move from side to side, not just up and down.

COCKROACHES GET THEIR...

sense of smell mainly from their antennae.

THERE ARE AROUND...

4,000 types of cockroach in the world.

SOME CAN...

hold their breath underwater for 40 minutes!

SOME HUMANS ARE...

allergic to cockroaches.

MOST COCKROACHES HAVE...

18 knees!

COCKROACHES EYES ARE MADE UP OF...

4,000 lenses, which help them see in all directions at the same time.

GLOSSARY

Abdomen – the largest part of an insect's three-part body; the abdomen contains most of the important organs.

Antennae – a pair of special sense organs found at the front of the head on most insects.

Arthropods – any bug that has jointed legs; insects and spiders are arthropods.

Bristles – short, strong hairs.

Camouflaged – when animals blend in with their surroundings so they can't be seen easily.

Carnivores – animals that only eat meat.

Decompose – when plants and animals break down into tiny pieces that can be used to help other life grow.

Digestive system – the organs that are used to process food.

Exoskeleton – a hard outer covering that protects and supports the bodies of some bugs.

Functional – in working order.

Herbivores – animals that only eat plants.

Instars – a stage between moults for a developing insect nymph.

Larvae – a wormlike creature that is the juvenile (young) stage in the life cycle of many insects.

Mate – one of a pair of animals that live or have babies together.

Microbe – a tiny living thing, so small that they can only been seen through a powerful microscope.

Moulting – the process of shedding the body's surface layer so that it can be replaced by a new one.

Nymphs – the juvenile (young) stage in the life cycle of insects that do not produce larvae.

Omnivores – animals that eat both plants and meat.

Pheromone – a scent substance produced by many kinds of animals that is used to communicate certain information or "messages".

Predators – animals that hunt and eat other animals.

Pronotum – a tough shield that protects the head of some cockroaches.

Pupation – the process by which insect larvae change their body shape to the adult form.

Receptors – tiny organs that detect things such as smell, heat, and vibration (see right).

Rotting – the process of decomposition by which the bodies of dead animals and plants are broken down.

Scavengers – animals that eat dead and rotting plants and animals.

Segment – a part of something that is divided into a number of similar parts.

Skeleton – an internal structure of bones that supports the bodies of large animals such as mammals, reptiles, and fish.

Species – a group of living things that share characteristics and features.

Springtail – a six-legged bug which is not considered a true insect.

Sprinting – running at top speed.

Sub-tropical – belonging to a region near Earth's equator where the climate is always warm.

Thorax – the middle part of an insect's body where the legs are attached.

Tropical – belonging to the region around the Earth's equator where the climate is always hot.

Vibrations – to move back and forth very quickly.

INDEX

A
abdomen 6, 19, 30
adults 19, 20
African cockroaches 26
American cockroaches 23, 25
animal food 8
antennae 7, 12-13, 21, 28, 30
arthropods 5, 21, 30

B
banana roach 26
brain 7, 28
breathing 19, 29
bristles 13, 30
bugs 7, 8, 16, 30

C
camouflaged 26
Cape Mountain cockroach 27
carnivores 8
clean-up insects 9

D
decomposers 9
digestive system 6, 9
droppings 9, 14, 23, 25

E
egg cases 18-19, 27, 30
exoskeleton 5, 20, 30
eyes 7, 12, 29

F
family groups 4
females 16-17, 31
finding houses 22, 26
flying 16-17
food 6, 8-9, 12-13, 14-15, 23
forests 4, 27

G
German cockroaches 23, 25
giant cockroaches 24, 27
green banana roach 26

H
head 5, 6-7, 30-31
herbivores 8
hexapods 7
hissing 24
houses 22, 26
humans 22-23, 29

I
insects 4-5, 6-7, 11, 14, 19, 20-21, 23, 30
instars 21, 30

K
knees 19, 29, 31

L
larvae 30
leaf cockroach 26
legs 5, 6-7, 11, 13
light 12

M
Madagascan cockroach 24
males 16, 31
mating 16, 18, 30
microbes 9, 25, 30
moulting 21, 30
mountain cockroach 27
mouth 7, 14, 28

N
night 4, 10,
nymphs 19, 20-21, 25, 27

O
omnivores 8, 30
ootheca 18-19, 27, 30
organic waste 22
Oriental cockroaches 23

P
parental care 25
pheromones 16-17
plants 20, 30-31
predators 8, 10, 24, 26,
pronotum 7, 31
pupation 19, 31

R
racing cockroaches 24-25
receptors 12-13, 31
rotting 8, 12, 14, 31
running 10-11

S
scavengers 5, 8, 31
segments 12, 21
shape variations 26-27
size variations 26-27
skeleton 5, 30-31
smell 12, 14-15, 28, 31
snakes 21
sound defence 24
spoiling food 14-15, 23
springtails 7, 31
sprinting 11, 31
sub-tropical regions 4, 31
Surinam roach 16-17

T
temperature 12-13
thorax 6, 31
tropical regions 4, 27, 31

U
understanding cockroaches 5

V
vibration receptors 12-13

W
walking 10
wings 5, 6, 16-17, 21, 30
wood roaches 9, 25